MIAOW!

Sir Patrick Moore
CBE FRS

Cats really are nicer than people!

www.hubbleandhattie.com

The Hubble & Hattie imprint was launched in 2009 and is named in memory of two very special Westies owned by Veloce's proprietors.
Since the first book, many more have been added to the list, all with the same underlying objective: to be of real benefit to the species they cover, at the same time promoting compassion, understanding and co-operation between all animals (including human ones!)

Hubble & Hattie is the home of a range of books that cover all-things animal, produced to the same high quality of content and presentation as our motoring books, and offering the same great value for money.

Thanks to my good friend, Chas Parker, for the idea behind this book

More titles from Hubble & Hattie

A Dog's Dinner (Paton-Ayre)
Animal Grief: How animals mourn (Alderton)
Bramble: the dog who wanted to live forever (Heritage)
Cat Speak: recognising & understanding behaviour (Rauth-Widmann)
Clever dog! Life lessons from the world's most successful animal (O'Meara)
Complete Dog Massage Manual, The – Gentle Dog Care (Robertson)
Dieting with my dog: one busy life, two full figures ... and unconditional love (Frezon)
Dinner with Rover: delicious, nutritious meals for you and your dog to share (Paton-Ayre)
Dog Cookies: healthy, allergen-free treat recipes for your dog (Schöps)
Dog-friendly Gardening: creating a safe haven for you and your dog (Bush)
Dog Games – stimulating play to entertain your dog and you (Blenski)
Dog Speak: recognising & understanding behaviour (Blenski)
Dogs on Wheels: travelling with your canine companion (Mort)
Emergency First Aid for dogs: at home and away (Bucksch)
Exercising your puppy: a gentle & natural approach – Gentle Dog Care (Robertson & Pope)
Fun and Games for Cats (Seidl)
Groomer's Bible, The The definitive guide to the science, practice and art of dog grooming for students and home groomers (Gould)
Know Your Dog – The guide to a beautiful relationship (Birmelin)
Life Skills for Puppies: laying the foundation for a loving, lasting relationship (Zulch & Mills)
Miaow! Cats really are nicer than people! (Moore)
My dog has arthritis – but lives life to the full! (Carrick)
My dog is blind – but lives life to the full! (Horsky)
My dog is deaf – but lives life to the full! (Willms)
My dog has hip dysplasia – but lives life to the full! (Häusler)
My dog has cruciate ligament injury – but lives life to the full! (Häusler)
Older Dog, Living with an – Gentle Dog Care (Alderton & Hall)
Partners – Everyday working dogs being heroes every day (Walton)
Smellorama – nose games for dogs (Theby)
Swim to recovery: canine hydrotherapy healing – Gentle Dog Care (Wong)
The Truth about Wolves and Dogs: dispelling the myths of dog training (Shelbourne)
Waggy Tails & Wheelchairs (Epp)
Walking the dog: motorway walks for drivers & dogs (Rees)
Walking the dog in France: motorway walks for drivers & dogs (Rees)
Winston ... the dog who changed my life (Klute)
You and Your Border Terrier – The Essential Guide (Alderton)
You and Your Cockapoo – The Essential Guide (Alderton)

Publisher's note

All of the illustrations used in this book have come from Sir Patrick's personal collection; they hang on his walls, and line his mantelpieces. As such, some are not of the best quality, regrettably, although each and every one is precious to him

First published in April 2012 by Veloce Publishing Limited, Veloce House, Parkway Farm Business Park, Middle Farm Way, Poundbury, Dorchester, Dorset, DT1 3AR, England. Fax 01305 250479/e-mail info@hubbleandhattie.com/web www.hubbleandhattie.com
ISBN: 978-1-845844-35-6 UPC: 6-36847-04435-0. © Sir Patrick Moore & Veloce Publishing Ltd 2012. All rights reserved. With the exception of quoting brief passages for the purpose of review, no part of this publication may be recorded, reproduced or transmitted by any means, including photocopying, without the written permission of Veloce Publishing Ltd. Throughout this book logos, model names and designations, etc, have been used for the purposes of identification, illustration and decoration. Such names are the property of the trademark holder as this is not an official publication.
Readers with ideas for books about animals, or animal-related topics, are invited to write to the editorial director of Veloce Publishing at the above address. British Library Cataloguing in Publication Data – A catalogue record for this book is available from the British Library.
Typesetting, design and page make-up all by Veloce Publishing Ltd on Apple Mac. Printed in India by Imprint Digital Ltd

Contents

MiAOW!

1940 and wartime: Flying Officer Patrick Moore. I was a bomber command navigator throughout the Second World War against Germany.

MiAOW!

This little book is about cats – cats of all shapes, sizes and colours. But as my own two two beloved cats, Jeannie and Ptolemy, have a starring role in it, perhaps I ought to start by saying something about them.

My old cat, Smudgie, had come to the end of her life, aged twenty. She had had a long and happy life, but I missed her terribly. Then I had to go into hospital for a knee operation (which didn't work!), and, whilst there, was shown a notice which read: 'Kittens for sale, £15.' At home once again, a friend went to investigate, and brought me a small, squeaking kitten, which was Jeannie. She was put into my arms, and that was that: love at first sight. Jeannie is black and white, and we established that, by birth and family, she is a Norwegian Forest Cat. Quite how she came from Norway to Bognor Regis, I know not!

After a while, we thought that Jeannie might like a companion. A cat belonging to a friend of mine had produced kittens, and one of these – jet-black with lovely green eyes – was brought

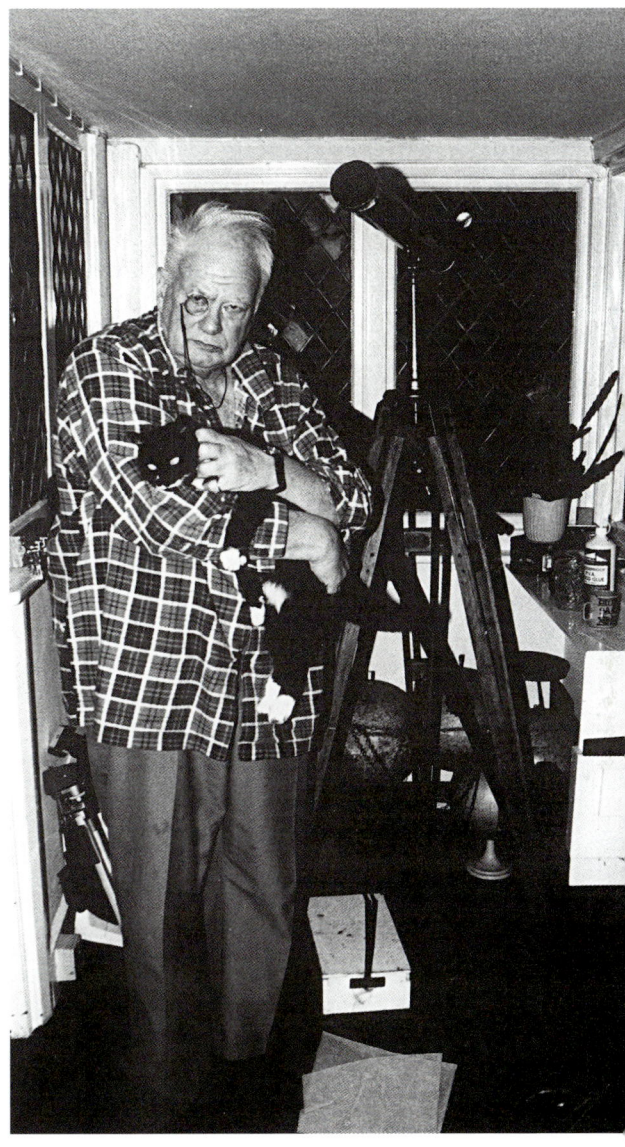

Here, I am holding my lovely Smudgie, who I had before the arrival of Jeannie and Ptolemy. She was a stray who came to me as a kitten and lived with me for twenty happy years.

My mother, Gertrude, a trained opera singer, was always very close to me, and always with me. In this picture she is holding our beloved ginger cat, Rufus, who was not only lovely to look at but also lovely by nature. Rufus had many happy years and when he finally didn't wake up, I was heartbroken.

to my house for me to see. Jeannie approved. The kitten looked around, fixed his eyes on me, and talked. I understood that what he was saying was: "I want to be your cat. I want to live here and make this my home. Do please adopt me!"

I stroked his head and he purred. "Your name is Ptolemy."

My family had increased by one.

MIAOW!

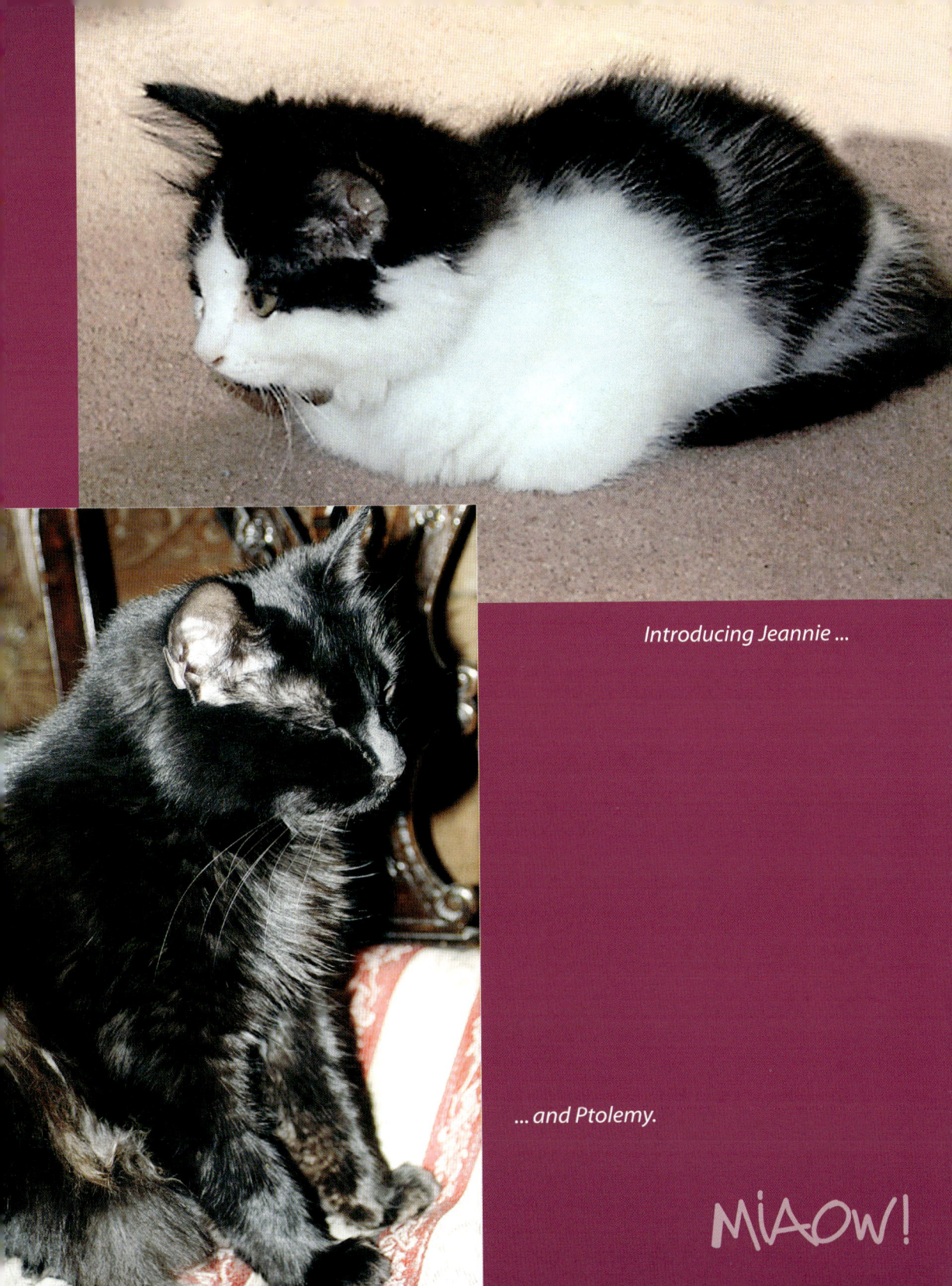

Introducing Jeannie ...

... and Ptolemy.

MIAOW!

Cats really are nicer than people!

I chose the title for my book quite deliberately. Obviously, it is dangerous to generalise, but given a choice between the average person and the average cat, I would opt for the cat any time, and in this book I will try and explain why.

Here goes!

My starting point is a cricket match, played some time ago between my club, Selsey, and our deadly rivals, West Wittering. It was a blazing hot August day, with the temperature well in the nineties (Fahrenheit, please note, not Centigrade; I have no use for what I call creeping metrication), and the match was very close. I claim that my quick leg-breaks bowled off a long, leaping run were decisive; I was well pleased with my final analysis of 8 for 69, but at the fall of the last Wittering wicket, I was decidedly relieved. I could not have gone on for much longer, and was as tired as if I had done a hard day's work (I imagine: not that I have ever worked in the conventional sense; I have always been far too busy).

What kind of reception would I receive upon my return home? From any human being, a comment about my 'untidy'

Here I am captaining our cricket team, during a match at Selsey ground, although I am purely a bowler and bat at number 11.

appearance, perhaps, together with a reminder that whilst I was 'relaxing' on the cricket field, there was a back lawn that badly needed mowing.

It so happened that there were no people in the house, just my two beloved cats, Jeannie and Ptolemy. As I arrived home, Jeannie jumped down from the hall chair on which she'd been

Mother and myself. We always stayed together and were absolutely devoted to each other. I could hardly believe it when she died at the age of 94, but at least she did so peacefully and without a long illness. I miss her terribly.

One could not ask for anything better by way of a welcome. People are never predictable, but cats are always the same, and I find it comforting to know that they will be there waiting for me.

So, all in all, they really are nicer than humans, although I appreciate that this may be regarded as too sweeping a generalisation because of the obvious language problem.

However, I pride myself that I have a pretty good idea of what is going on inside the little heads of my cats, and that I can speak basic Cat, though not so well as basic French, say, and we cannot carry on a conversation in the way that people can. Whether cats can converse fluently with each other I know not. I expect they can, but can't prove it, of course. Looking

resting, came up to me, squeaked, and asked for her ear to be tickled. Ptolemy was not far behind, emerging from my bedroom and greeting me with an affectionate 'miaow.'

My bed is a favourite resting place for my cats …

… where they can completely relax, at peace with the world, and invite a tummy tickle.

MIAOW!

at Jeannie I find it hard to believe that cats have unkind thoughts about each other, or, for that matter, about me. True, they live in a sheltered environment, and I have been accused of keeping "two very spoilt cats," but that is really beside the point. Find me the individual who does not ever harbour unkind thoughts, then step back and look for the halo above his or her head!

Whenever I come home, or when I am at home, I am always prepared for love and affection from one black cat and one black and white cat.

They never fail me.

MIAOW!

The man who hated cats

Strange to say, there are those who do not like cats, and who either ignore them or are downright hostile toward them. I find this most peculiar, but undoubtedly true.

One of my friends – I'll call him Jake – was one such person. Every time he came to my home, he studiously avoided Jeannie and Ptolemy. On one occasion, when Ptolemy appeared ready to jump onto his lap, he looked absolutely horrified. At first I thought that he might be allergic to cats, which would be unfortunate for him, but not his fault.

But this wasn't an allergy: this was sheer dislike.

Until one day.

I was relaxing in my study, when Jake appeared. I hadn't heard him come in because he has a key to my front door. There I was, cuddling Jeannie, and there was the Cat-Hater. What was going to happen?

What actually followed was totally unexpected.

As he looked at Jeannie and I, Jake's whole expression changed. He gingerly stepped forward and rubbed Jeannie's ear. Predictably, she purred.

This went on for some minutes

This notice is in the entrance to my house. Many people think it is a joke ... it isn't, of course.

as I watched, my mouth open in disbelief.

Jake looked up at me and grinned. "You know," he said, "she really is nice!" Far from being a temporary phase, as I suspected it might be, Jake quickly made friends with Ptolemy also (who is

a friend of all the world), and his conversion was complete. He and his wife live in the next village to me. They visited the local cat rescue centre and took home a kitten, which has now grown into a very handsome ivory-coloured cat called Zebbie (short for Zebra), because of his stripes.

What caused this sudden change of heart? I have to admit that I do not know, but when Jake comes to my house now, both cats make haste to give him an affectionate greeting.

I'll end this chapter by recounting a conversation I had with a

This sticker adorns a window in the hall of my home. Although I don't know where it came from, its sentiment is one I certainly agree with!

neighbour the other day. It went like this:

Him: "I rather like cats, but they kill birds."

Me: "That's their nature."

Him:"And they kill so many. How many have yours caught today?"

Me: "None."

Him: "That's because they have their own closed garden. All in all, thousands of birds are caught every week."

Me: "Quite so. By the way, did you enjoy that roast chicken you had last Sunday?"

Silence.

A good close-up of Jeannie. You will agree she has a lovely face, complete with little black patch on her chin.

MIAOW!

Can one choose a cat?

I have said that Jeannie and Ptolemy are 'my' cats, but is that really true? And, in any case, how do cats and people come together?

I well remember meeting my two.

My very beloved cat, Smudgie, had died at the age of almost twenty years, and I was heartbroken. She was a lovely cat, and I think as devoted to me as I was to her.

But we have to accept the fact that cats have shorter lives than ours, and there is nothing we can do about this. I simply had to keep in mind that I had given Smudgie almost twenty perfect years, and she had enjoyed every day of her life.

But there it was: did I search around for another cat, or did I not?

I was unsure, because I did not want to go through that heartbreak again.

However, my hand was more

or less forced when a close friend who was a house guest visited me in the hospital where I was having a knee operation.

"Something that might interest you;" he said, "there is an advertisement in the *Chichester Observer* from someone who has kittens for sale. Shall I go and have a look?"

I was not really feeling well enough to think properly about it, so simply said: "Well, go and have a look. It cannot do any harm."

He did just that, and came to see me again the next day. "Well, I had a look at them. There are five, and they tell me that the little one is called Jeannie. I just could not resist her. Therefore, we have now acquired a new member of the family!"

It was a couple of days before I was able to come home from hospital, but when I did I went directly to my dining room, where I was greeted by my friend with

Jeannie and Ptolemy kissing each other. Generally speaking, they get on really well and never fight, although sometimes they swear at each other!

something in a basket, which he passed to me. "This is Jeannie."

There are some moments that you will always remember, and, in my case, this was one. Jeannie put her little head on one side and looked up at me. She squeaked. I knew then there was absolutely no question – it was love at first sight. And so Jeannie became a much-loved member of my household, and as such she has remained.

The point I am making here is that there was never any doubt in my mind as soon as I set eyes on her.

She quickly settled down, and became very much a part of my life. But when she reached four years old I began to wonder whether she might like a feline companion. Please don't misunderstand me; she was absolutely settled in and at home, and clearly very fond of me and everyone else around her. But I have to admit that I cannot talk Cat fluently, and it might just be that she would like someone who could.

Alan, a great friend who lives opposite, came to see me one day, and told me about a friend of his who had a cat which had just

produced kittens. One of these kittens was pure black, and he wondered if I might like to have him?

Well, obviously, we had to have the consent and approval of everyone in the household, so

A typical Ptolemy picture: as you can see, he is very black indeed! The small patch of light which appears on his chest doesn't show up in this photo.

Alan brought the little kitten to my house in a basket. As soon as Alan placed the basket on the floor, the little kitten got out and jumped up at my legs. He put his head in my hands and talked to me. I could understand every word that the little fellow was saying: "I want this to be my home. I want to be with you. Please give me a home!"

Again, as with Jeannie, it was love at first sight. I stroked his

head. "Your name is Ptolemy; let me introduce you to Jeannie, and see what she thinks of you."

Had Jeannie and Ptolemy disliked each other on sight, the situation would have been incredibly awkward, and I simply do not know how it might have been resolved. Luckily, this was not the case as the two cats nestled up against each other, and that was that. Today, years, later, they are still very much together.

Can you now see what I mean about 'choosing' a cat? A person does not choose a cat, the cat chooses them. In my case there were no problems at all, and we remain a very happy and united household.

Jeannie and Ptolemy are under the impression that they have me precisely where they want me.

They have!

Jeannie had been named even before she came to me, but Ptolemy had not. So why did I choose that particular name?

Ptolemy takes his ease in the garden, just outside my study window.

It has an historical association because ancient Egypt was ruled by the Ptolemies, and in that country cats are sacred, which cats have never forgotten. But I have no Egyptian connections, and the name was chosen for quite another reason.

My maternal uncle was a barrister, but over one hundred years ago he gave up the law, went on the stage and became a well-known actor in comic opera. Uncle Reg had the role of Ptolemy in a London show called *Amasis*, which was set in Egypt. His parents had just acquired a black kitten which they called Ptolemy after the character that Reg was playing, and this is where the name came from.

MiAOW!

Feline intelligence

It is generally said that the cleverest of all animals are seals and their close relatives, followed by elephants, with dogs some way down the scale and cats even lower. I am quite sure that this is not correct, however, and I absolutely refuse to rank cats below dogs. Cats are far, far cleverer.

There are two immediate points here that I regard as important. First, it is wrong to confuse cleverness with cunning, although they may often seem the same.

One has only to look at the Front Bench and Cabinet members of our present House of Commons. They are certainly cunning, because they have the ability to make others believe what they say, but can you really think that they are actually clever?

The answer must be a resounding 'No!' Away from their cunning politics, it would be hard to find a bigger collection of dunces.

Now for my second point. It cannot be said that a cat's brain works in exactly the same way as a man's; it doesn't. There are fundamental differences, but we cannot pinpoint these differences precisely because of the language barrier. My brain does not work in the same way as that of an African Bushman, but given enough time we could make headway: I am confident that I could learn enough Bushmanese to hold a proper conversation, but I could never learn enough Cat, and the barrier here must remain.

Many of us have read Hugh Loftus' children's books featuring Dr Doolittle, and a great many of us have enjoyed Rudyard Kipling's *The Jungle Book*, but did Loftus and Kipling hit the proverbial nail squarely on the equally proverbial head? We cannot tell.

Many intelligence tests for cats have been devised, but, as with humans, a good score in such a test is not always indicative of a high IQ, because it is not easy to weed out the cunning percentage (Cun P).

Recently, I devised a test which I tried out on my two cats, Jeannie and Ptolemy.

As I expected, both did well: indeed, Jeannie would have been in the very top rank had she not

Ptolemy with my Woodstock. He has never tried to type: although he is a very clever cat, I don't think he could manage it …

incurred a penalty of ten marks for getting shut in the airing cupboard.

You may like to try out my test, overleaf, on your own cat, but first take a look at the sample question which follows.

• Your cat is sitting in the fireplace, close to the fire, which is getting hotter. Does he –

a *move out of the danger zone?*
b *stay where he is purring, even though clearly too hot?*
c *wait to be picked up and moved?*
d *go to sleep, oblivious of the heat?*

It is fairly apparent that the sensible answer in this case is (a), but not all cats will be sensible, and may take one of the other options.

Apply this test question to your cat and make your decision. Does he merit an (a)? He may not. Tick the letter that you think applies to him and be honest about it!

One of my beloved cats, who had a long and happy life, was as thick as several short planks, and if given this test would have ended up with a substantial minus score.

My answers are given at the end; you may not agree with all of them, but DON'T CHEAT by looking them up first!

Here, then, are the questions to put to your cat. Of course, they cannot answer directly themselves (although I can speak reasonable Cat, I am not good enough for this!), which means that we must answer for them, which is what I have done with my two.

Note, by the way, that throughout the test I refer to the cat as 'he.' This may be considered sexist but I do this only because I absolutely refuse to refer to a cat as 'it,' which I find frankly insulting. In this instance, 'he' is taken to mean either sex.

Good luck!

Question 1

If you call him, will he:
a look at you and blink sleepily?
b take absoutely no notice?
c jump up and wait to be cuddled?
d give a miaow to acknowledge you?
e walk over to you?

Question 2

Your cat is laying stretched out on his back, with his legs in the air.
You tickle his tummy; does he:
a yawn and miaow softly?
b jump up and attack your hand, claws out?
c as B but with claws sheathed?
d ignore you totally?
e get up and move somewhere else to sleep?

Question 3

You are lying comfortably asleep but your cat wants some attention.
To wake you, will he:
a sit on your head?
b Miaow loudly and continuously in your ear?
c wait patiently, purring softly?
d deliberately knock over an object, such as a coffee cup?
e let out a piercing screech?

Question 4

Does he use his litter tray when indoors:
a usually?
b always?
c sometimes?
d very seldom?
e only if strategically placed?

Ptolemy lies alongside a pen that was given to me around fifty years ago by a 'well-wisher.' I've never discovered who it was that so kindly sent it, so have never been able to thank them. It has resided on my desk all these years.

Question 5

If he has to take a pill, will he:

a disappear under the sofa?
b swallow the pill without fuss?
c fight like a wounded tiger?
d take the pill, but immediately spit it out?
e wail as if being attacked?

Question 6

If he is accidentally shut in a room or cupboard, will he:

a miaow softly until released?

b simply wait for rescue?

c try to open the door?

d miaow as loudly as possible?

e keep scratchng the door in the hope that he will be heard?

Question 7

You call him when he is within earshot. Does he:

a come straight to you?

b indicate that he has heard, but stay where he is?

c take no notice at all?

d purr to show that he has heard, then roll over onto his back to have his tummy tickled?

e pause, look at you, and then come to you?

Question 8

You have some very special treats in a bag, which he sees you place in a drawer. Will he:

a remember where they are?

b forget all about them?

c keep going to the drawer and looking at it?

d try to open the drawer?

e sit beside the drawer and squeak persistently?

Question 9

You have been away for a day or two. When you return, will he:

a greet you with real affection?

b look at you with casual interest?

c walk round you, sniffing?

d regard you with suspicion?

e beat a hasty retreat?

MIAOW!

Question 10

You are asleep, and he has no food or water down. Does he:

a jump on top of you?
b miaow loudly in your ear?
c wait patiently until you wake?
d pat your face with his paw?
e deliberately knock over an object, such as a table lamp?

Question 11

There is a loud bang nearby (such as a firework, or a dropped tray). Does he:

a jump up, miaow, and retreat?
b jump up, miaow, and stay where he is?
c take absolutely no notice?
d withdraw quickly and silently?
e squeak and look around to see what has made the noise?

Question 12

He has been outdoors, and comes in damp. Does he:

a rub against you until dry?
b make a beeline for the nearest fire/radiator?
c shake himself and shower everything within a yard?
d stay damp until he dries naturally?
e miaow loudly and persistently until you fetch a towel and dry him?

Question 13

He has climbed a tree and become stuck. Does he:

a wait placidly for rescue?
b squeak plaintively in the hope that someone will hear?
c attempt a very risky descent?
d miaow continuously as loudly as possible?
e stay absolutely still and quiet?

Answers overleaf ...

Answers

The correct answer in each case scores ten points. Incorrect answers do not score.

Question 1 – c
Question 2 – a
Question 3 – c
Question 4 – b
Question 5 – b
Question 6 – d
Question 7 – a
Question 8 – e
Question 9 – a
Question 10 – b
Question 11 – c
Question 12 – e
Question 13 – d

Results

100-130 marks
Highly intelligent and very affectionate. Ideal companion and friend.

60-90 marks
Great companion and average intelligence.

0-50 marks
Frankly, a dimwit – but no less lovable for that!

Heavenly felines

Jeannie in my study with what looks like a chunk of rock, but is, in actual fact, a meteorite that fell at Barwell in Leicestershire in 1986. I found the piece myself, and handed it into the museum, but was told to keep and display it myself, and leave it to the museum in my will. This I have done.

Look up at the sky on a dark, clear night, and you will see stars – thousands of them. In ancient times, these stars were grouped into 'constellations,' and given names. The names did not really mean anything because the stars in any particular constellation need not be associated with each other. The stars are at different distances from the earth, and we are dealing with nothing more significant than line-of-sight effects (electro-magnetic waves travelling in a straight line).

Remember, too, that the stars are a long way away. Our earth moves around the sun at a distance of 93,000,000 miles: we should appreciate that the sun is an ordinary star that looks much brighter and feels much hotter only because it is so close to us on the cosmical scale. All of the other stars are so far away that it is cumbersome to give the distances in miles, just as it would be awkward to give the distance in inches between, say, London and Manchester.

Luckily, a more convenient unit of

measurement is available to us. Light travels at 186,000 miles per second. In a year, therefore, it can cover almost six million million miles, a distance which is known as the light year. Even the nearest star beyond the sun is over four light years away.

Now consider two bright stars in the northern hemisphere of the sky, called by us Castor and Pollux (again, the names we give are meaningless: call the stars what you like). These two stars have been grouped in the same constellation, but we now know that Pollux is 32 light years away, and Castor 45 light years, so there is no actual association between the two. If we were observing from a different vantage point in space, these two stars could well be placed on opposite sides of the sky.

Now for the constellation names.

The first serious star-gazers were probably the Chinese and the Egyptians, both of whom had their own constellation patterns (we do not have full maps). Then came the Greeks and the birth of true astronomy.

Jeannie admires my orrery; a device that shows movements of the planets around the sun. This particular orrery was made in 1720.

Ptolemy paying close attention to the orrery, although I am not sure he knows what it's all about.

The last and greatest of the Greeks was Ptolemy, who lived around AD120, and who left us a magnificent book, known generally as the *Almagest*, which tells us most of what we know about ancient astronomy (actually, this is the Arabic title; the Greek original has not survived).

Ptolemy drew up a complete map of the sky, and named 48 constellations, all of which we still use today, albeit slightly modified. Most people will be aware of a number of Ptolemy's constellations: Ursa Major, Latin for Great Bear; Cassiopeia, the Queen; Cygnus, the Swan; and Orion, the Hunter, a truly splendid constellation that dominates the evening sky all through winter and early spring.

Dogs are well represented: Canis Major (the Great Dog) contains Sirius, the most brilliant star in the entire sky, while Canis Minor (the Little Dog) is led by the bright star Procyon, and Canis Venatici (the Hunting Dogs, Asterion and Chara) strain at the leash close to the tail of the Great Bear.

But cats there are none. True, there was a lion, and also a lynx, but these are not the same thing.

Though Ptolemy's constellations were accepted, there was a long period, several centuries ago, when other astronomers produced maps and introduced constellations of their own, few of which were generally accepted. Two of these new constellations were Felis, the Cat (put into the sky by Johann Bode in 1775), and Noctua, the Night Owl (by Burritt in 1833). Both of these groups adjoined Hydra, the Watersnake, which sprawls all the way from close to the celestial equator to the far south, and is actually the largest of Ptolemy's original 48 constellations.

Neither Felis or Noctua had any bright star or distinctive pattern, and both were soon forgotten. One has to admit that there was little justification for them, although I cannot help but feel a little sad at the demise of the Owl and the Pussycat ...

My worst review

I have been an author all my life, earning my living by writing, and, by now, I have produced over a hundred books, most of them on astronomy. Of course, the books have been reviewed, and, on the whole, the reviewers have been very kind. For many years, I have been happy with the reviews that my books received. Although I have had the occasional lukewarm comment, I had never had a bad review from anyone of note – until a little while ago.

To tell the whole story: I had begun work on a book called *Travellers in Space and Time*. After three days I had finished the first chapter and started on the second. Quite a number of words had been written with my usual fluency, but I could not shake the feeling that something was not quite right, though what that something was I could not say.

I read my typescript – yes, I know that typewriters were being phased out in favour of word processors at that time, but I had remained faithful to my ancient Woodstock which had served me well for so long – and my unease persisted. I had a couple of whiskies; same result, so I decided to get some sleep. I went off to bed telling myself that I was simply being super-critical.

By now it was one in the morning, and Jeannie and Ptolemy had settled into their usual routine. I should explain at this point that we have a very large, enclosed 'cat garden,' without exit to the outside world. It is accessed via a catflap in the music room, which is open at all times except during the hours of real darkness. Both cats are extremely well house-trained, and never perform their natural functions where they shouldn't. They're not even very keen on using litter trays, much preferring to go outside into their garden.

On this occasion I checked that neither cat wanted to go through the flap and then closed it for the night, satisfied that all was well.

Eventually dropping off to sleep, I didn't have a good night. As usual, Jeannie curled up at the foot of my bed, where she stayed all night. I didn't know where

![photograph]

For some reason, my study desk is a magnet: here, Ptolemy prepares to settle down as I look on.

Ptolemy was, though probably on the settee in my study.

When I awoke in the morning it seemed set to be a perfectly ordinary day. The sun was shining and the garden – mine and the cats' – looked lovely. I opened the catflap, then put the kettle on to make tea. I walked into the study and got the shock of my life.

As I have already said, both Jannie and Ptolemy are exceptionally clean cats, but on this occasion, Ptolemy had performed in no uncertain terms, not at random, but exactly on my typescript!

I looked at him and he looked back at me. Was there a knowing expression in those beautiful eyes?

Well, first things first. It was essential I clean up my typescript, which meant the use of gloves

Ptolemy in my arms. To the left you'll notice my ancient Woodstock typewriter, on which all of my books were written until last year. Using both middle fingers only, I could type at 90 words per minute on this machine. Today, I have to use a computer, on which I am not nearly as fast.

and a bucket. However, almost as soon as I began the job of

cleaning up I realised that it wasn't going to be possible. The entire

MIAOW!

typescript was sodden, but that wasn't the worst of it, if you see what I mean ... I tried to separate a couple of pages and iron them out with a complete lack of success. I hadn't saved or copied the work, so it was lost forever.

It's important to appreciate here that the trouble was confined solely to my papers, with not a scrap anywhere else. It was quite uncanny, and for a moment, I imagined that Ptolemy had read the work, hated it, and decided that it wasn't up to standard. Then reason prevailed; clever though he may be, his reading ability is certainly not up to that! There had to be another explanation ...

Ptolemy in my study 'assisting' witih the signing of the contract for this very book!

Ptolemy has never misbehaved since, and so the mystery remains, however.

Coincidence? Probably, though I suppose it is just possible that he saw me working the previous day, realised that I was dissatisfied, and decided to help. I will never know for sure.

There is a corollary to this episode.

The next day, I began to re-write my book. It came out quite different to the original version, and was much, much better.

So there we have it: Ptolemy did me my very worst review, but prevented me submitting a manuscript that was well below my usual standard. In the end, *Travellers in Time and Space* turned out to be one of my most successful books, so I suppose I have every reason to be grateful for Ptolemy's 'review.'

A narrow squeak

There are some things I don't like to think about. What follows is one.

It was evening, the sun had barely set, and it was time to check on the whereabouts of my feline friends. Ptolemy was easy – he was sprawled out in the garden – but of Jeannie there was no sign. I wasn't immediately worried because we have a fairly large house, and there are very many places where a little cat can go and be virtually out of sight.

As we hunted for Jeannie, John, a friend staying with me who is almost as fond of the two cats as I am, helped, but we simply could not find her. We searched every room in the house, checked in the garden – and began to become really worried.

Was it possible that Jeannie had got out of an open window, decided on a stroll in the big wide world, and now couldn't find her way home?

We continued our search for well over an hour, but Jeannie could not be found. We knew all of

A leisurely wash and brush up – a 'cat's lick' – for Jeannie, stretched out on my study floor.

her hiding places, and our search was extremely thorough. By this time, I really *was* worried.

We decided that if Jeannie had gotten outside she might be in the main garden, or even (heaven forbid!) the road. We hunted in the garden first, calling her name and looking behind bushes, then went out into the road. I felt quite sure that if Jeannie had gone outside she would be lost, frightened, and unable to find her way home. Remember, all her life had been spent in the house or the cat garden, and in the great wide world she would be completely unable to cope.

By now, both John and I were deeply concerned for Jeannie. John suggested we have one last look in the house, just in case we had somehow missed something.

We began downstairs and worked our way through my study, music room, slide room, and all of the other rooms, but of Jeannie there was still no sign. We then went upstairs and made another search of the bedrooms, though by now I was feeling extremely pessimistic.

Then came a welcome sound. In one of the spare bedrooms we heard what sounded like a faint squeak. John and I looked at each other; where was the sound coming from? Undoubtedly it was above us, but this in itself posed a problem as one thing that my home, Farthings, lacks is a proper loft. There is a loft of sorts, but this is very small, and before long turns into a narrow passage leading I know not where.

Entry into the loft space is via a pull-down ladder in a bedroom. John – much more agile than I – climbed the ladder into the main 'loft' and then quietly waited.

There! There it was again, a faint, pitiful squeak, which seemed to come from somewhere in the long, narrow passage. As John peered into the darkness he saw a pair of green eyes looking back at him. Jeannie had been found but was in real danger, as the passage ended in a hole which was the gap between the outer and inner walls (Farthings is very old and had been built that way). Jeannie was sitting at the end of the relatively safe part of the passage; two or three feet further on and she would have plunged into the hole, and I had no idea how we

MIAOW!

Jeannie, engaging in conversation with a visiting rabbit: they seemed to get along very well.

would get her out. Probably it would mean making a hole in the wall, and in a 13th century house, this is not a good idea.

John squirmed along the passage as far as he could; frankly, it was so cramped that it was very dangerous for him. Jeannie was a finger's width out of reach, and so frightened that she seemed rooted to the spot. With one last effort John stretched out one arm (the other was captive underneath his prone body) and grabbed ... thank goodness Jeannie made no attempt to back away from him or she may have fallen backward into the abyss.

The battle was not yet over as John had only the one hand free with which to hold on to Jeannie. If she wriggled free, we were back to square one.

John began to draw himself back along the passage; luckily, Jeannie remained still as he brought her with him. In the bedroom below I waited anxiously, as scared as Jeannie.

At last came relief. John managed to get back into the loft space and stand up, taking Jeannie into his arms. Moments later he was at the top of the ladder, making his way down. I expect you can appreciate how very glad I was to see them!

Jeannie really was so frightened that she was trembling, and her usually immaculate coat was filthy with dust and cobwebs,

Like many cats, Jeannie loves to be cuddled, and here I am, obliging her.

which she can't have enjoyed at all. We brushed and combed her, then sat down to have a much needed whisky and soda.

It was raining gently outside, and Jeannie did something that she never ordinarily did: she went outside and came in wet. By the time we had dried her, most of the dust and dirt had gone, but she spent the next few hours washing herself, nevertheless.

If you have ever seen a truly terrified cat you will agree that it is not a pleasant sight. However, all's well that ended well, and at least Jeannie was not physically hurt. We tended her carefully, and at last she settled down to a peaceful sleep.

My worst fear throughout was that she may plunge into the wall cavity, which could so easily have happened. Once John and I had downed a few more whiskies, he went to the loft opening (which had been somehow left slightly open, allowing Jeannie access) and secured it so effectively that nothing would ever dislodge it.

By the morning, Jeannie was back to normal. I wonder if she knew just how worried John and I were during her rescue operation? At least now there is no chance of this happening again; although I had been so sure that there were no possible dangers in the house, I suppose I had become complacent.

It took an episode like this to make me realise just how much I love my cats.

MIAOW!

In the tree tops

I have already recounted the tale of when Jeannie went missing: really the only time in her life when she has done so. Ptolemy has, at times, been rather evasive, and has taken a good deal of tracking down, but he never usually goes far, and always turns up not long after the hunt for him has begun.

But there was one occasion when he didn't.

Late one afternoon I wanted to check on my cats, because visiting me was a family newly arrived in Selsey – mother, father, and teenage son, Adrian. Real 'cat people' who had expressed a wish to see Jeannie and Ptolemy.

Jeannie was in full view in the garden, but of Ptolemy there was no sign, and the usual search failed to locate him. I cannot say that at this point I was deeply worried, because I know he has an uncanny habit of turning up safe and sound, nonchalantly strolling into view when he feels like it. But, after a couple of hours without him putting in an appearance, I did begin to feel a little apprehensive. All of us made

A typical pose by Ptolemy. He likes cameras and is always ready to be photographed.

a careful search of the house and garden, but no Ptolemy.

It was a warm evening and we sat down at a table outside the front door and poured drinks. Suddenly, Adrian grabbed me by the arm.

"Look! There he is, up the tree." We do have some tall trees in the main garden, but I had never given them much thought. I didn't even know what kind of trees they were – ash, oak, or something else – but, frankly, I wouldn't have cared if they proved to be monkeypuzzles. They did have considerable branches, though, which extended more or less over the garden, and it was among these that Adrian was pointing.

I have heard many stories about cats that climb trees, can't then get down, and have to be rescued by the fire brigade, but this particular branch wasn't at the tree top, and appeared to be within reach of somebody with a long ladder. Adrian volunteered to be the 'somebody,' so we fetched the ladder from my garage and leant it against the tree.

The cat moved up to the next branch.

Adrian climbed the ladder but, at the top, discovered he wasn't very keen on heights. Neither am I, for that matter, but felt that I ought to at least show willing. I gripped the ladder sides

He also likes getting into mischief ... here he is playing hide-and-seek in a cardboard box containing packaging.

MiAOW!

I think this is a good close-up of Ptolemy. He has such lovely eyes that shine in the dark. Whenever we look for him in the garden after nightfall, we can always spot him because of his glowing eyes.

and began to climb, slowly and cautiously, upward. I felt the ladder rungs beneath my feet bend ominously; it wasn't really strong enough to take a man's weight. Risking life and limb, I got to the top of the ladder, but my quarry was out of reach. I descended the ladder, pretty discouraged.

What to do? I knew some neighbours had a much longer ladder, which they used for roof repairs. I went to them and put my plea. They were totally understanding, and helped me to convey the ladder from their garden to mine. Surely this would be long (and strong) enough? We put it against the tree, and it appeared pleasingly robust.

Clearly, someone had to go up the ladder and see if they could grab Ptolemy, and equally clearly it had to be me who did it, so up I went. As I climbed higher and higher I became less and less comfortable, but to turn back would be the act of a coward. Oh dear ...

Suddenly, there was a most unexpected development; there was Ptolemy, sitting on the lawn, looking faintly surprised at all the commotion. Thankful that I need climb no higher, I made my way down and stepped off the ladder. I approached Ptolemy to reassure myself it really was him; it was.

So, what had happened? I can only say that, from ground level, it is very hard to distinguish a cat from a squirrel, and there, in the tree, sat a very large squirrel, which I swear was laughing at me!

So, the very next time that you can't find your cat, before you get out the ladder or ring the fire brigade, make absolutely sure that your quarry is not just a squirrel whose home is in the tree tops.

Actually, we do have a squirrel drey just outside my study, in the main garden. There are two inhabitants, who I have called Sidney and Sabena. They have never met either cat, and it is most unlikely that they ever will, so whether they would be friendly or hostile is something I shall never discover.

Cat garden

This is a rather serious chapter, and I do hope that it will make you think.

There are many dangers facing cats who live in the great wide world: dogs, traffic, foxes, infectious diseases, and so on. For Jeannie and Ptolemy, I have eliminated these risks by giving them their own large, enclosed garden that is still open to sunlight and rain, though the latter does not matter as the cats can get into the house via the cat flap. (Ptolemy will sometimes go out and get wet; Jeannie never will!)

Feral cats are natural wanderers who would not tolerate any restriction of movement, and they must take their chances in the outdoors. Many domestic cats will also wander, if given the

The special 'cat garden.' Both Jeannie and Ptolemy love it here. With plenty of room in which to exercise, they are also completely safe.

chance, and they present a totally different problem, because, in most cases, they have not been taught the skills necessary to survive on their own.

Hence my garden for Jeannie and Ptolemy.

Opinion on this point differs, of course, but, frankly, I doubt whether any cats could be happier than my two.

MIAOW!

Left: Jeannie carefully studying the sunshine recorder in her garden ...

Ptolemy in typical restful pose in the garden he shares with Jeannie ...

... who relaxes inside on the carpet after a nice lunch.

Left: ... and finding the sunniest spot in which to relax.

MiAOW!

Epilogue

As I write these words Ptolemy is sitting on my desk, purring, and Jeannie is curled up beside me.

Just as it should be.

Jeannie and Ptolemy find enough space for both of them on my study desk.

Looks like Ptolemy's been at the champagne, if the cork he's cuddling is anything to go by!

MiAOW!

I love the way that the shadows in this picture make Jeannie look like a tiger on the prowl ...

... although in the next instant, she's looking as innocent as a kitten!

MIAOW!

Dear Jeannie

Very soon after this book was written, sadly, my beloved Jeannie was taken ill with kidney failure, and nothing could be done for her.

I had no choice but to have her put to sleep. Had she lived on, even a little while longer, she might have suffered. As things were, she had no pain or suffering at all, and slipped quietly away.

Some people may not understand how it is possible to have such a deep love for a little cat, but my love for Jeannie went very deep indeed.

When I held her for the very last time and kissed her, I did not say 'Goodbye, Jeannie' because I know she will wait for me, and I will see her again. So instead I said 'Au revoir, Jeannie, dear, until we next meet.'

At least I know she had 13 very happy years. Everybody loved her and everybody misses her – Ptolemy included. So this book is dedicated to Jeannie.

A lovely close-up of Jeannie, who is clearly concentrating very hard on something or other.

Jeannie, suitably framed, in a painting that was a gift to me. This is such a nice painting of Jeannie that I thought it worth framing. It hangs in my study.

MiAOW!

128 pages; 191 colour pictures; £14.99*

Visit our website (www.hubbleandhattie.com) or call (01305 260068) for full details. P&P extra. *Subject to change

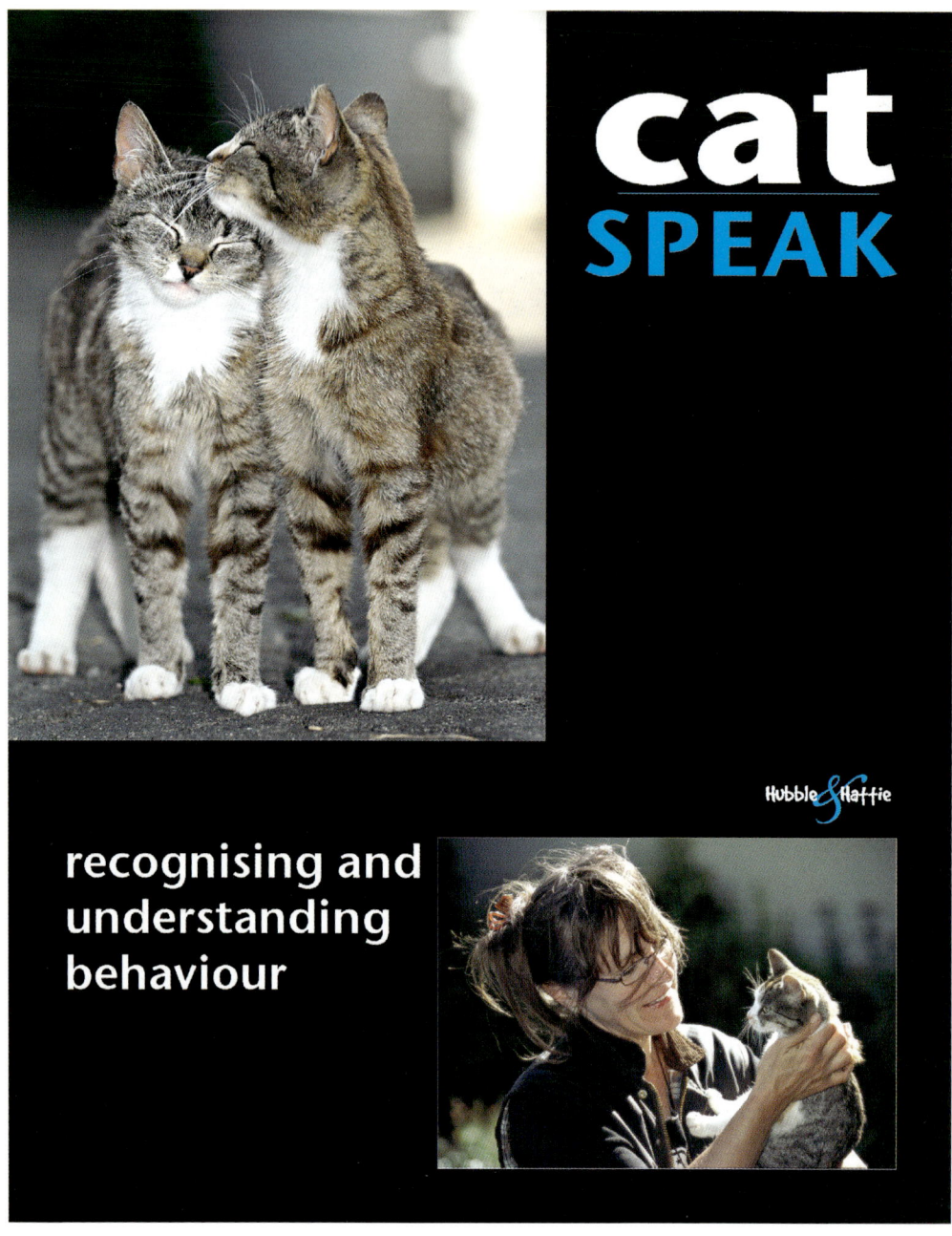